Copyrights by Roy Richard Sawyer 2011-2021
All rights reserve
ISBN-10: **1546319123**
ISBN-13: **978-1546319122**

The Easiest Way to Understand Algebra

Roy Richard Sawyer

Table of Contents

INTRODUCTION	1
HOW TO SOLVE THE EQUATIONS	1
Practice 1. Solving the equations.	3
Practice 2. Solving the equations.	4
Practice 3. Solving the equations	8
Practice 4. Solving the equations.	9
Practice 5. Solving the equations.	9
SYSTEM OF EQUATIONS	**10**
Practice 6. Solving system of equations.	16
QUADRATIC EQUATIONS	**17**
Practice 7. Solving Quadratic Equations	19
Logarithmic equations	**19**
A Decimal Logarithm	19
Practice 8: logarithmic equations	26
APPENDIX 1. ANSWERS TO EQUATIONS	**27**
Practice 1 Answers.	27
Practice 2 Answers.	27
Practice 3 Answers.	28
Practice 4 Answers	28
Practice 5 Answers.	28
Practice 6 Answers.	28
Practice 7 Answers.	29
Practice 8 Answers.	29

APPENDIX 2. SOLUTIONS.   30

Practice 1 solutions   30
Practice 2. Solutions.   32
Practice 3. Solutions.   35
Practice 4. Solutions.   38
Practice 5. Solutions.   40
Practice 6. Solutions.   42
Practice 6. Graph solutions.   47
Practice 7. Solutions.   57
Practice 8 solutions   64

My eBooks on Amazon.com   **71**
ABOUT THE AUTHOR   **72**

INTRODUCTION

There is a voice inside you that always says," You can't do that. Leave it." This is not true, but you believe this voice because you get some benefit from accepting the advice. You get a good excuse to slip away from your work. You can go out with friends, watch TV, or do whatever you want. However, you cannot feel genuine pleasure from your entertainment because there is a part of you who wants to be proud of your achievements. This voice tells to "Keep trying! You can do that! You are smart." Believe in what this voice is saying! This tutorial will show you a way of thinking that will help you to understand math. This tutorial is for anyone who wants to feel comfortable using a mathematical formula; who wants to comprehend the beauty of algebraic expressions.
Did you ever feel frustration looking at your math textbook? Forget it! Fall in love with math!

HOW TO SOLVE THE EQUATIONS

Math is a delightful field of activity. Having just a pen and a piece of paper, you can invent whatever you want. You can wander around the paper with numbers and symbols caring about just one thing: equality should be equality, nothing more. Let us imagine that you are the first great mathematician. People are only familiar with arithmetic: how to add, subtract, multiply, and divide. In school, they study boring things such as these expressions:
$2 + 3 = 5$ or $7 - 4 = 3$
You are the first who suspects there is a way to express a common idea of the equations written above.
First you write: $a + b = c$ or $c - a = b$. Now you have discovered common rules that can help people to solve any equation. To verify the discovery, you have to perform experiments with numbers.
Let us write a simple equation: $4 + 8 = 12$
Let's add any number to the left side of the equation.
$4 + 8 + 3 = 12$
What did you get?
$15 = 12$
This is incorrect! How can you fix your equation? You must add the same number to the right side of the equation.
$4 + 8 + 3 = 12 + 3$. What did you get? $15 = 15$

You discovered the first rule for equations. This rule says: "If you add the same number to the left side and to the right side of an equation, this equation will still be true." To express this rule in a common way you can write:

If $a + b = c$ then $a + b + n = C + n$ where a, b, c, n equal any numbers.

Are you a genius? Of course, you are! Go ahead. Let us try another experiment.

What happens if you subtract any number from the left side of an equation?

$5+2=7$

$5+2-5=7$

What did you get? $2 = 7$

This is incorrect, but you know how to fix your equation. You must subtract the same number from the left and right side of the equation.

$5 + 2 - 5 = 7 - 5$. Then $2 = 2$.

Congratulations! You discovered the second rule for equations. This rule says: "If you subtract the same numbers from the left and the right side of an equation, this equation will still be true."

Or you can write:

if $a + b = c$ then $a + b - n = c - n$ where a, b, c, n equal any numbers.

What other kinds of experiments can you do? You can multiply one side of an equation by some number. Let's write an equation:

$5 - 1 = 4$

What happens, if you multiply the left side of the equation by 7?

$(5 - 1)7 = 4$ then $28 = 4$

This is not true. Try to multiply both sides of the equation by 7.

$(5 -1)7= 4 \times 7$. Then $28 = 28$

You discovered one more rule for equations. The third rule says:

"If you multiply the left and the right side of an equation by the same number, this equation will still be true."

if $a - b = c$ then $(a - b) n = (c)n$

One more question. What happens if you divide one half of an equation by any number?

$4 + 6 = 10$

$(4 + 6)/2 = 10$

then $5 = 10$.

You can ask yourself, "How many times will I make the same mistake?" But you have the knowledge to fix the problem.

You must divide both sides of the equation by the same number.

$(4 + 6)/ 2 = 10/ 2$ then $5 = 5$

You discovered the fourth rule for equations. This rule says:

"If you divide the left and the right side of an equation by the same number the equation still will be true."

So, you can write:

The Easiest Way to Understand Algebra

if $a + b = c$ then $(a + b)/n = c/n$

Where a, b, c, are any numbers, but n does not equal 0 because you can't divide numbers by 0.

People will ask you, "What kind of benefit can you get from these rules?" Your response will be, "You can use these rules to solve any equation."

Let us write an equation where one number is unknown.

$X - 3 = 11$

How can we solve this equation? Try to apply the first rule:

If you add the same number to the left and the right side of an equation, this equation will be true.

For our equation, it is convenient to add 3 to both sides of the equation.

$X - 3 + 3 = 11 + 3$

Since $-3 + 3 = 0$

Then $X = 11 + 3$

So, $X = 14$

Let us try to solve an equation where all numbers are represented by letters.

$X - b = c$

Apply the first rule to solve this equation

$X - b + b = c + b$

Since, $-b + b = 0$, then

$X = c + b$.

To solve the equation $X + b = c$, we can apply the second rule.

If $X + b = c$ then

$X + b - b = c - b$

$X = c - b$.

The next example is $X + 7 = 15$

then $X + 7 - 7 = 15 - 7$ and $X = 8$

Do not proceed until you perform some exercises to become comfortable using the first few rules of equations.

Practice 1. Solving the equations.

Solve for X:
1. $X - 5 = 0$
2. $X + 11 = 3$
3. $X - ab = 4$
4. $X - Y = Z$
5. $X - 2a = c$
6. $X + 3ab = bc$
7. $X + k = 1 + k$
8. $X - ab = a - ab$
9. $X + c = c - b$
10. $X - 2a = a - ab$

11. X + cb = 3cb - c
12. X - 5 + a = 2a - 5
13. X + 3 - k = 6 - 3k
14. X - 1 - ab = ab - 1
15. X - a - b = a - b
16. X + 2a - 3c = 3a - 2c

You can find the answers in appendix 1. If your answer is wrong try again. If you can't get the right answer, read the solution in appendix 2.
Let's solve the equation
4X - 5 = 15
You can apply the first rule.
4X - 5 + 5 = 15 + 5 then 4X = 20.
How can you find X? You can apply the fourth rule.
If you divide both sides of an equation by the same numbers, this equation will still be true.
4X / 4 = 20 / 4, then X = 5.
To solve equation
aX - b = c
Apply the first rule.
aX - b + b = c + b,
then aX = c + b
Now apply the fourth rule.
If aX = c + b, then aX / a = (c + b) / a
and X = (c + b)/ a
Do not read any more until you perform some exercises.

Practice 2. Solving the equations.

Solve for X:
1. 2X - 3 = 5
2. 3X - 5 = 4
3. 5X + 6 = 36
4. 8X - 5 = 43
5. 7X - 2 = 19
6. 4X + 8 = 20
7. 6X - a = 2a
8. 2X + b = 13b
9. 7X + 3a = a + b
10. 4X - 2a = 4 + 2a
11. 4X - 3a = a
12. 3X - 2b = 6 - 14b
13. 6X - 2a = 24b - 20a
14. aX - 3a = ab - 2a

The Easiest Way to Understand Algebra

15. 2aX + ab = 2a - ab
16. 3aX - c = 3ac - 7c

Answers are in appendix 1.
Solutions are in appendix 2.
If you have such an equation to solve:
X/a - 5 = 6
Then apply the first rule:
X/a - 5 + 5 = 6 + 5
X/a = 6 + 5
X/a = 11
Then apply the third rule.
X/a * a = 11 *a
X = 11a
Let's solve the equation:
2X - 4b = 2bc
Apply the first rule:
2X - 4b + 4b = 2bc + 4b,
then 2X = 2bc + 4b
Apply the third rule:
2X/ 2 = (2bc + 4b) / 2
You should know how to divide a binomial by a monomial.
If you have forgotten it, you could find the rule by yourself. Can you write
(2bc+ 4b)/2 = 2bc/2 + 4b/2? Yes, you can.
Let us check. Suppose c = 2 and b = 3.
To divide a binomial by 2, try to divide each monomial by 2
2*3*2/2 + 4*3/2 = 12
Now try to solve the binomial first and then divide by 2
(2*3*2 + 4*3)/2 then 24/2 = 12
We got the same answer. It means that
(a+ b)/2 = a/2 + b/2.
We discovered a rule: To divide a binomial by a number, divide each monomial
inside the binomial by that number. Come back to your equation.
2X = 2bc + 4b. Then
2X / 2 = 2bc / 2 + 4b / 2
Then X = bc + 2b
You can factor out b and get
X = b (c + 2)
Whenever you don't know the rule, you can put any numbers in place of
the letters and check equality. Discover rules by yourself.
Let us solve a more complicated equation:

$$\frac{5X - 5}{5X} = 10$$

Multiply both sides of the equation by 5X.

$$\frac{5X - 5}{5X} = 10$$

5X - 5 = 50X

Use the 2nd rule, subtract 5X from both sides:

5X − 5 − 5X = 50X − 5X

-5 = 45X

or 45X = - 5

Divide both sides by 45

45X/45 = -5/45

X = - 1/9

The following equation:

$$\frac{aX + b}{a + b} - X = c$$

Find the common denominator:

$$\frac{aX + b}{(a + b)} - \frac{X(a + b)}{(a + b)} = c$$

$$\frac{aX + b - X(a + b)}{(a + b)} = c$$

$$\frac{aX + b - aX - bX}{a + b} = c$$

Since + aX - aX = 0, our equation becomes simple:

$$\frac{b - bX}{a + b} = c$$

Use the 3rd rule: multiply both sides of the equation by (a + b)

The Easiest Way to Understand Algebra

$$\frac{(b - bX)(a + b)}{a + b} = c(a + b)$$

Then $b - bX = c(a + b)$
Apply the 2nd rule, subtract b from both sides of the equation:

$b - bX - b = c(a + b) - b$

Then $-bX = c(a + b) - b$ Divide both sides by b:

$$\frac{-bX}{-b} = \frac{ac + bc - b}{-b} \qquad X = \frac{ac + bc - b}{-b}$$

To make this algebraic expression more beautiful, multiply the numerator and denominator by (-1).
You can do that because $(-1)/(-1) = 1$. If you multiply any number by 1, the number will not be changed.

$$X = \frac{(ac + bc - b)(-1)}{(-b)(-1)}$$

Then

$$X = \frac{b - ac - bc}{b}$$

The following equation:
$-2X = a - b$
It is not convenient for you to have a minus in front of 2X.
You can change the equation into a more convenient form.
Let's multiply both sides of the equation by -1

$(-2X)(-1) = (a - b)(-1)$ then you get

$2X = -a + b$ or $2X = b - a$

Divide both sides by 2:

$$\frac{2X}{2} = \frac{b - a}{2} \qquad X = \frac{b - a}{2}$$

There is another way to solve this equation:
$-2X = a - b$
Let's divide both sides of the equation by -2

7

$$\frac{-2X}{-2} = \frac{a-b}{-2} \quad X = \frac{a-b}{-2}$$

To make your result more beautiful, you can multiply the numerator and the denominator by – 1.

$$X = \frac{(a-b)(-1)}{(-2)(-1)} \quad X = \frac{-a+b}{2} = \frac{b-a}{2}$$

The following equation:
3a - 6X = 6X - 9a
You can see that on the left side of the equation you have -6X
And on the right side, you have +6X. It is more convenient for you to have a + in front of X. Therefore, you leave +6X on the right side and get rid of -6X on the left side of the equation.
Add 6X to both sides of the equation:
3a - 6X + 6X = 6X - 9a + 6X then 3a = 12X - 9a
Add 9a to both sides of the equation:

3a + 9a = 12X - 9a + 9a then 12a = 12X

X = a

Do not read any more until you perform some exercises.

Practice 3. Solving the equations

(Solve for X)
1. 1 - X = 5 - a
2. 1 - 2X = X - 4
3. a - 3X = b - X
4. 2a - 4X = 2X - 4a
5. 4b - 2X = 2X - 4b
6. ab + aX = 2aX + ac
7. ab + aX = 2aX – ac

Let us continue and discuss the equation: aX - bX = a – b
Factor out X which is a common factor for binomial aX - bX, then you will get X (a - b) = a - b
Divide each part of the equation by a – b

X (a - b)/ (a - b) = (a - b)/ (a - b)

$$X = \frac{a-b}{a-b}$$
X = 1
Do not read any more until you perform some exercises.

Practice 4. Solving the equations.

(Solve for X)
1. bX - 2b = aX - 2a
2. b - 2bX = a - 2aX
3. aX - bX = 1
4. aX - bX - cX = 2a - 2b - 2c
5. 3abX - 5a = 3acX + 13a
6. aX - bX = ac - bc
7. 9a - 4X = 5a - 2X
8. X - aX = 2 - 2a
9. aX - bX = b - a

Let us continue and solve the equation: aX - bX = 2b - 2a
Factor out X from the left side of the equation.
X (a - b) = 2b - 2a
Factor out 2 from the right side of the equation.
X (a - b) = 2(b - a) then divide both sides by (a - b)
$$\frac{X(a-b)}{a-b} = \frac{2(b-a)}{a-b}$$
Factor out (-1) from the numerator
$$X = \frac{2(-b+a)(-1)}{a-b} = \frac{-2(a-b)}{a-b} = -2$$

Or you can simplify this algebraic expression by factoring out (-1) from the denominator

$$X = \frac{2(b-a)}{(-1)(-a+b)} = \frac{2(b-a)}{-1(b-a)} = -2$$

Practice 5. Solving the equations.

Solve for X.
1. 5aX - 5bX = 10b - 10a
2. aX - bX - cX = c + b - a

9

3. $2X - 3aX = 6a - 4$
4. $3aX - 9bX = 27b - 9a$
5. $4bX - cX = 8c - 32b$
6. $abX - acX = ac - ab$
7. $X/2 - aX = 1 - 2a$
8. $aX/5 + 2a = 5a - 4aX$

You can find answers in appendix 1 and solutions in appendix 2.

SYSTEM OF EQUATIONS

Look at the equation $X + Y = 3$. X and Y are unknown. You can´t find neither X nor Y from this equation. You need additional information about the "relationship" between them. Such information may be included in an additional equation. For example: $X - Y = -1$. Now you have a system of 2 equations:
1. $X + Y = 3$
   $X - Y = -1$
There are several ways to solve it. The first way: solve for X in any equation, for example, the first one.
In order to do that, subtract Y from each side of the equation:
$X + Y - Y = 3 - Y$; Find X
$X = 3 - Y$
Then put (3 - Y) in place of X in the second equation ($X - Y = -1$).
You will obtain: $3 - Y - Y = -1$ or $3 - 2Y = -1$
Now solve that equation for Y. Add 2Y to ~~the~~ both sides of the equation.
$3 - 2Y + 2Y = -1 + 2Y$
$3 = -1 + 2Y$
Add 1 to the both sides of the equation.
$3 + 1 = -1 + 1 + 2Y$
$4 = 2Y$
Switch 4 and 2Y
$2Y = 4$
Divide both sides of the equation by 2.
$2Y/2 = 4/2$
$Y = 2$. Now put 2 in place of Y in any original equation to find X.
One of the original equations is $X + Y = 3$
$X + 2 = 3$;
Subtract 2 from both sides of the equation.
$X + 2 - 2 = 3 - 2$;
$X = 1$
$Y = 2$

The second way:
1. $X + Y = 3$
   $X - Y = -1$
You can sum the left sides of both equations and sum the right sides of both equations.
$X + Y = 3$ and $X - Y = -1$ Then

$(X + Y) + (X - Y) = 3 + (-1)$

Or you can write it in this way:

$$+\begin{array}{r} X + Y = 3 \\ X - Y = -1 \\ \hline 2X + 0 = 2 \end{array}$$

Then $X = 1$
Put 1 in place of X in any equation.
$1 + Y = 3$
Subtract 1 from both sides of the equation.
$1 + Y - 1 = 3 - 1$
$Y = 3 - 1$
$Y = 2$

We can solve system equations using their graphs. If we plot each equation, we will get two straight lines. The point of the intersection of the lines will have values X and Y that fit both equations.
To draw a graph for an equation, we have to change it to a general form:
$Y = aX + b$
Let's start from the first equation: $X + Y = 3$
Subtract X from both sides of the equation
$X - X + Y = 3 - X$
$Y = 3 - X$
Find 2 points to draw the first equation. Assign any value to X and calculate the value of Y.
$X = 3$; $Y(3) = 3 - 3 = 0$
$X = 6$; $Y(6) = 3 - 6 = -3$
Two points are enough to draw a straight line.
Let's find points for the second equation. Change the equation form to a general one.
$X - Y = -1$
Subtract X from both sides of the equation.
$X - X - Y = -1 - X$
$-Y = -1 - X$
Multiply both sides by -1

-Y (-1) = (-1) (-1) - X (-1)
Y = X + 1
Find 2 points to draw the second equation. Assign any value to X and calculate the value of Y.
X = 5; Y (5) = 5 + 1 = 6
X = -5; Y (-5) = - 5 + 1 = -4
Now we can draw graphs for both lines.
In graph 1, you see that the lines intersection point has X = 1 and Y = 2.
They are the same values we found before.

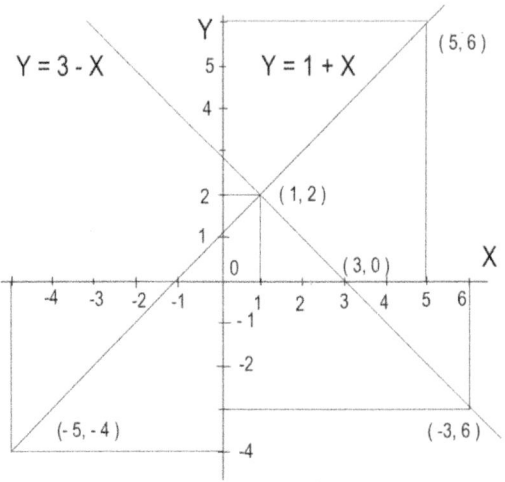

Graph 1. A point of the intersection: X=1, Y=2.

The next system of equations is:
2. 2X + Y = 5
    X + Y = 2
In this case, we have + in front of X and Y in both equations.
To eliminate one unknown member of the equation you have to subtract the second equation from the first one.
(2X + Y) - (X + Y) = 5 – 2 = 3

```
  2X + Y = 5
-  X + Y = 2
   X     = 3
```

Then put 3 in place of X in any original equation.

$3 + Y = 2$
Subtract 3 from both sides of the equation.
$3 + Y - 3 = 2 - 3$
$Y = 2 - 3 = -1$
$Y = -1$
Answers: $X = 3$; $Y = -1$;

Let's solve these equations in a graph.
The first equation is $2X + Y = 5$.
Change it to a general form $Y = aX + b$
Subtract 2X from both sides of the equation.
$2X - 2X + Y = 5 - 2X$
$Y = 5 - 2X$

Find 2 points to draw the first equation. Assign any value to X and calculate the value of Y.
$X = 0$; $Y(0) = 5 - 2*0$; $Y(0) = 5$
$X = 4$; $Y(4) = 5 - 2*4 = 5 - 8 = -3$
Let's find 2 points for the second equation.
$X + Y = 2$

Change it to a general form $Y = aX + b$
$X - X + Y = 2 - X$;
$Y = 2 - X$

Find 2 points to draw the second equation. Assign any value to X and calculate the value of Y.

$X = 4$; $Y(4) = 2 - 4 = -2$
$X = -4$; $Y(-4) = 2 - (-4) = 2 + 4 = 6$

In graph 2, we can see that the point of the intersection is (3, -1). The same answer we found before: $X = 3$, $Y = -1$

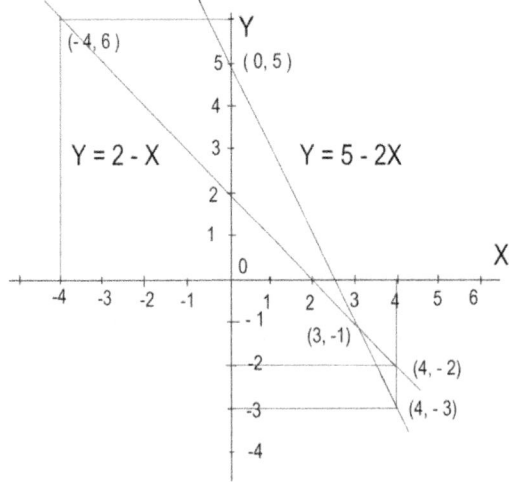

Graph 2. The point of the intersection: X=3, Y=-1.

The next system of equation is:
3. X - Y = 3
3X - 2Y = 4

To eliminate one unknown member of the equation, you have to get the same number in front of X in the first and in the second equation or the same number in front of Y in the first and in the second equation. In our case, in front of Y you have -1 in the first equation and -2 in the second equation.
To get -2Y in the first equation, you should multiply both sides of the equation by two. Then you will get:

2X - 2Y = 6
3X - 2Y = 4

Now subtract the second equation from the first one.

$$\begin{array}{r} 2X - 2Y = 6 \\ 3X - 2Y = 4 \\ \hline -X + 0 = 2 \end{array}$$

(with a minus sign before the first equation)

Multiply both sides of the equation -X = 2 by -1
(-X)(-1) = 2(-1)

then X = - 2

Put -2 in place of X in any original equation.
- 2 - Y = 3
Add 2 to both sides of the equation
- 2 - Y + 2 = 3 + 2
- Y = 5
Multiply both sides of the equation by - 1
- Y * (-1) = 5 * (-1)
Y = - 5
You can check your result. Put -2 in place of X and -5 in place of Y in both equations:
Check the first equation:
1. X - Y = 3
-2 - (- 5) = 3
-2 + 5 = 3
3 = 3
Check the second equation:
2. 3X - 2Y = 4
3*(- 2) - 2*(- 5) = 4
- 6 + 10 = 4
4 = 4
Let's solve these equations using graphs.
The first equation is X - Y = 3
Change it to a general form:
X - X - Y = 3 - X
- Y = 3 - X multiply both sides by -1
Y = X – 3
Find 2 points to draw the first equation. Assign any value to X and calculate the value of Y.
X = 0; Y (0) = 0 – 3 = - 3
X = 3; Y (3) = 3 - 3 = 0
The second equation 3X - 2Y = 4
Change it to a general form: subtract 3X from both sides.
3X - 3X - 2Y = 4 - 3X
-2Y = 4 - 3X
Divide each member of equation by -2
-2Y/ (-2) = 4/ (-2) - 3X/ (-2)
Y = - 2 + 3X/2
Switch the order of the members on the right side of the equation
Y = 3X/2 – 2

Find 2 points to draw the second equation. Assign any value to X and calculate the value of Y.

X = 4; Y (4) = 3*4/2 - 2 = 6 - 2 = 4
X = 2; Y (2) = 3*2/2 - 2 = 3 - 2 = 1

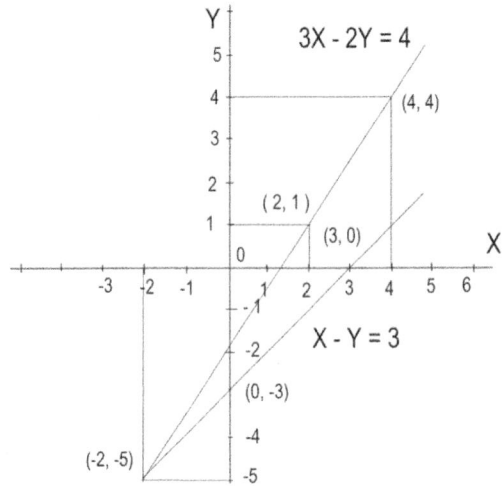

Graph3. The point of the intersection: X=-2, Y= -5.

Practice 6. Solving system of equations.

1. $X + Y = 1$
   $X - Y = -5$

2. $X + Y = 1$
   $2X - 2Y = 6$

3. $X + 2Y = 1$
   $2X + Y = -4$

4. $2X + Y = 5$
   $X + 3Y = 0$

5. $3X - Y = 5$
   $4X + 2Y = 10$

6. $4X + 2Y = 10$
   $4X - 2Y = 6$

7. $X - Y = 1.5$

7X + 2Y = 6

8. 3X + Y = 9
   X - 3Y = 3

9. 5X - 2Y = -7
   X + 3Y = 2

10. 9X + 3Y = 12
    X - 2Y = -1

QUADRATIC EQUATIONS

Let us examine the following equation:
$6X^2 + 3X - 3 = 0$
Express the left side of the equation as a product of two binomials. To perform that, factor the first and the last term of the equation. Factor $6X^2$. You can express $6X^2$ as $6X * X$ or $3X * 2X$
Then factor -3. You can express -3 as 1(-3) or 3(-1)
$6X^2 = (2X)(3X)$
Factor the last term $-3 = (-1)(+3)$ Combine them together: You will get

$6X^2 + 3X - 3 = (2X - 1)(3X + 3)$
because $6X^2 + 3X - 3 = 2X * 3X + 2X * 3 - 1 * 3X - 3$

You may check whether the expression on the left side of the equation equals to the expression on the right side.
Is $6X^2 + 3X - 3 = (2X - 1)(3X + 3)$?
Multiply $(2X - 1)$ by $(3X + 3)$ and you will get
$6X^2 + 6X - 3X - 3$ or $6X^2 + 3X - 3$
The two expressions are equal. Now you can write
$(2X - 1)(3X + 3) = 0$
If the product of two binomials equals 0 then each of them may be equal to 0.
$(2X - 1) = 0$ $(3X + 3) = 0$

Solve the first equation:
Add 1 to both sides of the equation

$2X - 1 + 1 = 0 + 1$
$2X = 1$
Divide both sides of the equation by 2:

$2X/2 = 1/2$ then $X = 0.5$

Solve the second equation: $(3X + 3) = 0$
Subtract -3 from both sides of the equation.
$3X + 3 - 3 = 0 - 3$
$3X = -3$
Divide both sides of the equation by 3:
$3X/3 = -3/3$
and $X = -1$
To check your solution, put each result in the original equation.
The original equation is: $6X^2 + 3X - 3 = 0$
$6 * (0.5)(0.5) + 3(0.5) - 3 = 0$
$1.5 + 1.5 - 3 = 0$
$3 - 3 = 0$
You can solve the same equation in a different way.
Factor $6X^2$ as $6X * X$
Factor the last term and you will get
$(6X - 3)(X + 1) = 0$
Now you can check:
$(6X - 3)(X + 1) = 6X^2 + 6X - 3X - 3 =$
$= 6X^2 + 3X - 3$
You got the original equation. Let us solve it.
$(6X - 3) = 0$
Add 3 to both sides of the equation.
$6X - 3 + 3 = 0 + 3$
$6X = 3$
Divide each side of the equation by 6
$6X/6 = 3/6$
$X = 0.5$
since $(6X - 3)(X + 1) = 0$
$(X + 1) = 0$
Solve that equation by subtracting 1 from both sides of the equation
$X + 1 - 1 = 0 - 1$
$X = -1$
You get the same results.
We can solve this equation with a quadratic formula:

$X = [-b +/- \operatorname{sqrt}(b^2 - 4ac)] / 2a$
$6X^2 + 3X - 3 = 0$
$a = 6$ (a number in front of X)
$b = 3$ (a number in front of Y)
$c = -3$ (a number without X or Y)

$X = [-b +/- \text{sqrt}(b^2 - 4ac)]/2a$
$X = [-3 +/- \text{sqrt}(3^2 - 4*6*(-3))]/2*6$
$X = [-3 +/- \text{sqrt}(9 - (-72)]/12$
$X = [-3 +/- \text{sqrt}(9 + 72)]/12$
$X = [-3 +/- \text{sqrt}(81)]/12$
$X = (-3 +/- 9)/12$
$X = (-3 - 9)/12 = -12/12 = 1$
$X = (-3 + 9)/12 = 6/12 = 0.5$
We got the same result: X = 1 and X = 0.5

Practice 7. Solving Quadratic Equations

1. $3X^2 - 75 = 0$
2. $2X^2 - 9X + 4 = 0$
3. $3X^2 - 5X - 2 = 0$
4. $4X^2 - 13X + 3 = 0$
5. $7X^2 - 29X + 4 = 0$
6. $5X^2 - 28X + 15 = 0$
7. $18X^2 + 12X - 6 = 0$
8. $3X^2 - 12X + 9 = 0$
9. $24X^2 + 55X - 24 = 0$
10. $12X^2 - 45X - 12 = 0$
11. $8X^2 - 28X + 12 = 0$
12. $12X^2 - 48 = 0$
13. $2X^2 - 11X + 9 = 0$
14. $14X^2 - 27X + 9 = 0$
15. $5X^2 - 37X + 14 = 0$

Logarithmic equations

A Decimal Logarithm

The meaning of logarithms and how to understand them. Let's look at this example.

$$\log_{10} 100 = 2$$

In our example 100 is an argument of the logarithm. 10 is a base of the logarithm and 2 is an exponent of the logarithm.
Logarithms of base 10 are known as the decimal logarithms.

How can you get 2 using 100 and 10? You can try performing all possible calculations. For example, 100 – 10 =90. It is not 2. 100 + 10 is not 2. 100/10 is not 2. What is left?

10 ^ 2 = 100

It means that the logarithm of an argument is the power by which the base must be raised to produce the argument. The logarithm of 100 equals 2 because to produce it from base 10 we must raise 10 by the power of 2.

Then, the decimal logarithm of 1000 will be equal to 3 because to get 1000 we must raise 10 by the power of 3.

It is said that the base of an algorithm cannot be equal to 1 or 0. Let's see why.

Imagine the following example. By which power must you raise 1 to get 100?

$$\log_1 100 =$$

It is obvious that 1 * 1 * 1 * 1 * 1 * 1= 1. You can use any power and you will not be able to get 100.

The same with zero. 0 * 0 = 0

A logarithm can be equal to 0 because 10 raised by the power of 0 = 1.

$$\log_{10} 1 = 0$$

Since any number raised by the power of 0 equals 1, the logarithm 1 of any base equals 0. This rule is called **The Zero Exponent Rule**.

Let's see with another example.

$$\log_{10} 100000 = 5$$

You can imagine 10000 as 100 * 100. Then

$$\log_{10} 100 \cdot 1000 = 5$$

You already know that Logarithm of 100 equals 2 and logarithm 1000 equals 3. Then you can write:

$$\log_{10} 100 \cdot 1000 = \log_{10} 100 + \log_{10} 1000$$

Or in common form

$$\log a \cdot b = \log a + \log b$$

Or

$$\log ab = \log a + \log b$$

You discovered a rule that is called The Product rule:

The logarithm of a product of two numbers equals the sum of the logarithm of the first number and the logarithm of the second number.

Now look at this example

$$\log_{10} 100 = 2$$

You can write 100 as 100000/1000. Then

$$\log_{10} 100000/1000 = 2$$

You know that logarithm 100000 = 5 and logarithm 1000 = 3. 5 − 3 = 2 then

$$\log_{10} 100000/1000 = \log_{10} 100000 - \log_{10} 1000$$

Or in common form:

$$\log a/b = \log a - \log b$$

You discovered a rule that is called The Quotient rule:

The logarithm of quotient of two numbers equals the difference of the logarithm of the numerator and the logarithm of the denominator.

Let's look at this example

$$\log_{10} 10000 = 4$$

You can write

$$10000 = 100^2$$

Then substitute 10000 in the logarithm with 100^2.

$$\log_{10} 100^2 = 4$$

You can write 4 as 2 * 2 then

$$\log_{10} 10000 = 2 \cdot 2$$

You know that logarithm of 100 equals 2. Substitute the one 2 with the logarithm of 100.

$$\log_{10} 100^2 = 2 \cdot \log_{10} 100$$

Or in common form:

$$\log N^m = m \log N$$

You discovered the Power Rule.

Next example:

$$\log_{10} 1/100 = -2$$

Because 10 raised by the power of -2 = 1/100

You can write -2 as 2 * (- 1).

You know that the decimal logarithm of 100 equals 2.

Then you can write:

$$\log_{10} 1/100 = -1 \cdot \log_{10} 100$$

Or

$$\log_{10} 1/100 = -\log_{10} 100$$

Or in common form:

$$\log 1/N = -\log N$$

You discovered the Reciprocal Rule.

Look at this example:

$$10^{\log_{10} 100} = 100$$

The decimal logarithm of 100 equals 2. The 10 raised by the power of 2 equals 100.

$$10^{\log_{10} 1000} = 1000$$

The decimal logarithm of 1000 equals 3. The 10 raised by the power of 3 equals 1000.

You discovered one more rule:

$$10^{\log_{10} N} = N$$

Let's discuss the following example.

$$100 = \log_{10} 10^{100}$$

To get 10 in power of 100, you have to raise 10 to the power of 100.

You can represent it in common form

$$n = \log_m m^n$$

Then any number n can be represented as a logarithm of the number m raised to the power of number n with the base number m.

Natural logarithms are the logarithms with base equals to Euler constant 'e'. The 'e' constant approximately equals to 2.718281828.

All the above rules apply to the natural logarithms as well.

Let's try to solve logarithmic equations. For the solution use formula

$$\text{If } \log_a f(x) = \log_a a^b \text{ then } f(x) = a^b$$

Where f(x) is some function of x.

1. $\log_{10} 2x = 2$

You can represent 2 as the logarithm of 10 raised to the power of 2 with the base of 10.

$$\log_{10} 2x = \log_{10} 10^2$$

Then 2x = 10^2.

2x = 100

2x / 2 = 100 / 2

X = 50

You can check if the solution is correct.

$$\log_{10} 2 \cdot 50 = 2$$

The solution is correct. The decimal logarithm of 100 equals 2.

2. $\log_{10}(2x-4) = 4$

You can represent 4 as the logarithm of 10 raised to the power of 4 with the base of 10.

$$4 = \log_{10} 10^4 \quad \text{then}$$

$$\log_{10}(2x-4) = \log_{10} 10^4 \quad \text{then}$$

$$2x - 4 = 10^4$$

Then 2x - 4 + 4 = 10000 + 4

2x = 1004

X = 502

3. $\log_{10} 6x + 1 = \log_{10} 2x + 3$

Let's move logarithms with x to the left and 1 to the right.

$$\log_{10} 6x - \log_{10} 2x = 3 - 1$$

Since log6x − log2x = log(6x / 2x), you can write

$$\log_{10} \frac{6x}{2x} = 2$$

Now you can represent 2 as logarithm 10 raised by the power of 2.

$$\log_{10} \frac{6x}{2x} = \log_{10} 10^2$$

Then 6x / 2x = 10^2

3x = 100   X = 100/3

4. $\log_{10} 100 = X$

Rewrite the equation in exponential form.

$$10^x = 100$$

Since 10 * 10 = 100, X = 2

5. $(x - 4)\log_{10} 10 = \log_{10} 100$

Divide both sides by logarithm of 10.

$$\frac{(x-4)\log_{10} 10}{\log_{10} 10} = \frac{\log_{10} 100}{\log_{10} 10}$$

The decimal logarithm of 100 equals 2 and the decimal logarithm of 10 equals 1.

X - 4 = 2 / 1 = 2
    x - 4 + 4 = 2 + 4
    x = 6

Practice 8: The logarithmic equations

1. $\log_{10} 3x = -1$

2. $\log_{10} 3x = \log_{10} 6x + 2$

3. $\log_{10} 10^{x+10} = \log_{10} 100$

4. $\log_{10}(x^2 - 2x + 2) = 1$

5. $\log_{10} 1/3x = 2$

6. $\log_{10} 40x + 1 = \log_{10} 4x + 3x$

7. $\log_{10} 4x = 2 - \log_{10} x$

8. $\log_{10} x^{(x+1)} = x + 1$

9. $\log_{10} 2x^2 = 2 + \log_{10} x$

10. $\log_{10} x = 3 - \log_{10} x^2$

APPENDIX 1. ANSWERS TO EQUATIONS

Practice 1 Answers.

1) X=5;
2) X= - 8;
3) X=4+ab;
4) X= Z+Y;
5) X= C+2a;
6) X=bc-3ab;
7) X=1;
8) X=a;
9) X=- b;
10) X = 3a - ab or X = a(3 - b);
11) X= c(2b -1);
12) X=a;
13) X=3-2k;
14) X=2ab;
15) X=2a;
16) X=a + c;

Practice 2 Answers.

1) X=4;
2) X=3;
3) X=6;
4) X=6;
5) X=3;
6) X=3;
7) X=a/2;
8) X=6b;
9) X=(b-2a)/7;
10) X=1+a;
11) X=a;
12) X=2-4b;
13) X= 4b - 3a;
14) X=b + 1;
15) X=1-b;
16) X=c - 2c/a;

Practice 3 Answers.

1) X=a-4;
2) X=5/3;
3) X= (a - b)/2;
4) X=a;
5) X=2b;
6) X= b-c;
7) X= b + c

Practice 4 Answers.

1) X= 2;
2) X= 1/2;
3) X= 1/(a-b);
4) X=2;
5) X= 6/(b-c);
6) X=c;
7) X=2a;
8) X=2;
9) X=-1;

Practice 5 Answers.

1) X= -2;
2) X= -1;
3) X= -2;
4) X= -3;
5) X= - 8;
6) X= - 1;
7) X=2;
8) X=5/7;

Practice 6 Answers.

1) X= - 2; Y=3;
2) X=2; Y=-1;
3) X=-3; Y=2;
4) X=3; Y=-1;
5) X=2; Y=1;
6) X=2; Y=1;
7) X=1; Y= - 0.5;

8) X=3; Y=0;
9) X=-1; Y=1;
10) X=1; Y=1;

Practice 7 Answers.

1. X = -5 or X = 5
2. X = 0.5 or X = 4
3. X = -1/3 or X = 2
4. X = 0.25 or X = 3
5. X = 1/7 or X = 4
6. X = 3/5 or X = 5
7. X = -1 or X = 1/3
8. X = 1 or X = 3
9. X = 3/8 or X =-8/3
10. X = -0.25 or X = 4
11. X = 0.5 or X = 3
12. X = 4 or X = 0
13. X = 4.5 or X = 1
14. X = 1.5 or X = 3/7
15. X = 0.4 or X = 7

Practice 8 Answers.

1. X = 1/30
2. X = 200
3. X = 2 or X = -8
4. X = 4 or X = -2
5. X = 1/300
6. X = 2/3
7. X = 5 or X = -5
8. X = 10
9. X = 50
10. X = 10

APPENDIX 2. SOLUTIONS.

Practice 1 solutions.

1. $X - 5 = 0$
Add 5 to both sides of the equation
$X - 5 + 5 = 0 + 5$
$X = 0 + 5$
$X = 5$

2. $X + 11 = 3$
Subtract 11 from both sides of the equation:
$X + 11 - 11 = 3 - 11$
$X = 3 - 11$
$X = -8$

3. $X - ab = 4$
Add ab to both sides of the equation:
$X - ab + ab = 4 + ab$
$X = 4 + ab$

4. $X - Y = Z$
Add Y to both sides of the equation:
$X - Y + Y = Z + Y$
$X = Z + Y$

5. $X - 2a = c$
Add 2a to both sides of the equation:
$X - 2a + 2a = c + 2a$
$X = c + 2a$

6. $X + 3ab = bc$

Subtract 3ab from both sides of the equation:

$X + 3ab - 3ab = bc - 3ab$
$X = bc - 3ab$

7. $X + k = 1 + k$
Subtract k from both sides of the equation:
$X + k - k = 1 + k - k$
$X = 1$

8. X - ab = a - ab
Add ab to both sides of the equation:
X -ab + ab = a - ab + ab
X = a - ab + ab
X = a

9. X + c = c - b
Subtract c from both sides of the equation:
X + c - c = c - b - c
X = - b

10. X - 2a = a - ab
Add 2a to both sides of the equation:
X - 2a + 2a = a - ab + 2a
a + 2a add up to 3a
X = 3a - ab or
X = a (3 - b)

11. X + cb = 3cb - c
Subtract cb from both sides of the equation:
X + cb - cb = 3cb - c - cb
X = 3cb - c − cb
X = 2cb - c
X = c (2b -1)

12. X - 5 + a = 2a- 5
Add 5 to both sides of the equation:
X - 5 + 5 + a = 2a - 5 + 5
X + a = 2a
Subtract a from both sides of the equation:
X + a - a = 2a - a
X = a

13. X + 3 - k = 6 - 3k
Subtract 3 from both sides of the equation:
X + 3 - 3 - k = 6 -3k - 3
X - k = 3 - 3k
Add k to both sides of the equation:
X - k + k = 3 - 3k + k
X = 3 - 2k

14. X - 1 - ab = ab - 1
Add 1 to both sides of the equation:

X - 1 + 1 - ab = ab - 1 + 1;
X - ab = ab
Add ab to both sides of the equation:
X - ab + ab = ab + ab
X = 2 ab

15. X - a - b = a - b
Add a to both sides of the equation:
X - a + a - b = a - b + a
X - b = 2a - b
Add b to both sides of the equation:
X - b + b = 2a - b + b
X = 2a

16. X + 2a - 3c = 3a - 2c
Subtract 2a from both sides of the equation:
X + 2a - 2a - 3c = 3a - 2c - 2a.
X - 3c = a - 2c
Add 3c to both sides of the equation:
X - 3c + 3c = a - 2c + 3c
X = a + c

Practice 2. Solutions.

1. 2X - 3 = 5
Add 3 to both sides of the equation:
2X - 3 + 3 = 5 + 3
2X = 8
Divide both sides of the equation by 2:
2X/2 = 8/2
X=4

2. 3X - 5 = 4
Add 5 to both sides of the equation:
3X - 5 + 5 = 4 + 5
3X = 9
Divide both sides of the equation by 3:
3X/3 = 9/3
X=3

3. 5X + 6 = 36
Subtract 6 from both sides of the equation:
5X + 6 - 6 = 36 - 6

The Easiest Way to Understand Algebra

5X = 30
Divide both sides of the equation by 5:
5X/5 = 30/5
X=6

4. 8X - 5 = 43
Add 5 to both sides of the equation:
8X - 5 + 5 = 43 + 5
8X = 48
Divide both sides of the equation by 8:
8X/8 = 48/8
X=6

5. 7X - 2 = 19
Add 2 to both sides of the equation:
7X - 2 + 2 = 19 + 2
7X = 21
Divide both sides of the equation by 7:
7X/7 = 21/7
X=3

6. 4X + 8 = 20
Subtract 8 from both sides of the equation:
4X + 8 - 8 = 20 - 8
4X = 12
Divide both sides of the equation by 4:
4X/4 = 12/4
X=3

7. 6X - a = 2a
Add a to both sides of the equation:
6X - a + a = 2a + a
6X = 3a
Divide both sides of the equation by 6:
6X/6 = 3a/6
X = a / 2

8. 2X + b = 13b
Subtract b from both sides of the equation:
2X + b - b = 13b - b
2X = 12b
Divide both sides of the equation by 2:
2X/2 = 12b/2

$X = 6b$

9. $7X + 3a = a + b$
Subtract 3a from both sides of the equation:
$7X + 3a - 3a = a + b - 3a$
$7X = b - 2a$
Divide both sides of the equation by 7:
$7X/7 = (b - 2a)/7$
$X = (b - 2a)/ 7$

10. $4X - 2a = 4 + 2a$
Add 2a to both sides of the equation:
$4X - 2a + 2a = 4 + 2a + 2a$
$4X = 4 + 4a$
Divide both sides of the equation by 4:
$4X/4 = 4/4 + 4a/4$
$X = 1 + a$

11. $4X - 3a = a$
Add 3a to both sides of the equation:
$4X - 3a + 3a = a + 3a$
$4X = 4a$
Divide both sides of the equation by 4:
$4X/4 = 4a/4$
$X = a$

12. $3X - 2b = 6 - 14b$
Add 2b to both sides of the equation:
$3X - 2b + 2b = 6 - 14b + 2b$
$3X = 6 - 12b$
Divide both sides of the equation by 3:
$3X/3 = (6 - 12b)/3$
$3X/3 = 6/3 - 12b/3$
$X = 2 - 4b$

13. $6X - 2a = 24b - 20a$
Add 2a to both sides of the equation:
$6X - 2a + 2a = 24b - 20a + 2a$
$6X = 24b - 18a$
Divide both sides of the equation by 6:
$6X/6 = (24b - 18a)/6$
$6X/6 = 24b/6 - 18a/6$
$X = 4b - 3a$

14. aX - 3a = ab - 2a
Add 3a to both sides of the equation:
aX - 3a + 3a = ab - 2a + 3a
aX = ab + a
Factor a out of term: ab + a
aX = a (b + 1)
Divide both sides of the equation by a:
aX/a = a (b + 1)/a
a should not be equal to 0!
X = b + 1

15. 2aX + ab = 2a - ab
Subtract ab from both sides of the equation:
2aX + ab - ab = 2a - ab - ab
2aX = 2a - 2ab
2aX = 2a (1 - b)
Divide both sides of the equation by 2a:
2aX/2a = 2a (1 - b)/2a
X = 1 - b

16. 3aX - c = 3ac - 7c
Add c to both sides of the equation:
3aX - c + c = 3ac - 7c + c
3aX = 3ac - 6c
Divide both sides of the equation by 3a:
3aX/3a = (3ac - 6c)/3a
3aX/3a = 3ac/3a - 6c/3a
X = c - 2c/a

Practice 3. Solutions.

1. 1 - X = 5 - a
Subtract 1 from both sides of the equation:
1 - X - 1 = 5 - a - 1
-X = 4 - a
Multiply both sides of the equation by (-1):
(-X) (-1) = (-1) (4 - a)
X = (- 4 + a)
X = a – 4

2. 1 - 2X = X - 4;
Subtract 1 from both sides of the equation:

1 - 2X - 1 = X - 4 - 1
- 2X = X – 5
Subtract X from both sides of the equation:
- 2X - X = X - 5 - X
- 3X = - 5
Divide both sides of the equation by -3:
- 3X/(-3) = - 5/(-3)
X = 5/3

3. a - 3X = b - X
Subtract a from both sides of the equation:
a - 3X - a = b - X - a
-3X = b - X - a
Add X to both sides of the equation:
-3X + X = b - X - a + X
-2X = b - a
Multiply both sides of the equation by (-1):
-2X (-1) = (-1) (b - a)
2X = a - b
Divide both sides of the equation by 2:
2X/2 = (a - b)/2
X = (a - b) / 2

4. 2a - 4X = 2X - 4a
Subtract 2a from both sides of the equation:
2a - 4X - 2a = 2X - 4a - 2a
- 4X = 2X - 6a
Subtract 2X from both sides of the equation:
-4X - 2X = 2X - 2X - 6a
-6X = - 6a
Multiply both sides of the equation by (-1):
(-6X) (-1) = (- 6a) (-1)
6X = 6a
Divide both sides of the equation by 6:
6X/6 = 6a/6
X = a

5. 4b - 2X = 2X - 4b
Add 2X to both sides of the equation:
4b - 2X + 2X = 2X - 4b + 2X
4b = 4X - 4b
Add 4b to both sides of the equation:
4b + 4b = 4X - 4b + 4b

$8b = 4X$
Divide both sides of the equation by 4:
$8b/4 = 4X/4$
$2b = X$
$X = 2b$

6. $ab + aX = 2aX + ac$
Subtract aX from both sides of the equation:
$ab + aX - aX = 2aX + ac - aX$
$ab = 2aX - aX + ac$
$ab = aX + ac$
Subtract ac from both sides of the equation:
$ab - ac = aX + ac - ac$
$ab - ac = aX$
Factor a out of the term: $ab - ac$
$a(b - c) = aX$
$aX = a(b - c)$
Divide both sides of the equation by a:
$aX/a = a(b - c)/a$
Where a is not equal to 0!
$X = a(b - c)/a$
$X = b - c$

7. $ab + aX = 2aX - ac$
Subtract ab from both sides of the equation:
$ab + aX - ab = 2aX - ac - ab$
$aX = 2aX - ac - ab$
Subtract aX from both sides of the equation:
$aX - aX = 2aX - ac - ab - aX$
$0 = 2aX - ac - ab - aX$
$0 = aX - ac - ab$
Add $ac + ab$ to both sides of the equation:
$0 + ac + ab = aX - ac - ab + ac + ab$
$ac + ab = aX$
Factor a out of the term: $ac + ab$
$a(c + b) = aX$
Divide both sides of the equation by a:
$a(b + c)/a = aX/a$
Where a is not equal to 0!
$b + c = X$
$X = b + c$

Practice 4. Solutions.

1. $bX - 2b = aX - 2a$
Add 2b to both sides of the equation:
$bX - 2b + 2b = aX - 2a + 2b$
$bX = aX - 2a + 2b$
Subtract aX from both sides of the equation:
$bX - aX = aX - 2a + 2b - aX$
$bX - aX = 2b - 2a$
Factor X out of the term: $bX - aX$
$X(b-a) = 2(b - a)$
Divide both sides of the equation by (b - a):
$X(b-a)/ (b - a) = 2(b - a)/ (b - a)$
Where (b-a) does not equal 0!
$X = 2(b-a)/(b-a) = 2$

2. $b - 2bX = a - 2aX$
Subtract b from both sides of the equation:
$b - b - 2bX = a - b - 2aX$
$-2bX = a - b - 2aX$
Add 2aX to both sides of the equation:
$-2bX + 2aX = a - b - 2aX + 2aX$
$-2bX + 2aX = a - b$
$2aX - 2bX = a - b$
Factor 2X out of the term: $2aX - 2bX$
$2X (a - b) = (a - b)$
Divide both sides of the equation by (a - b):
$2X (a - b)/ (a - b) = (a - b)/ (a - b)$
$2X = 1$
Divide both sides of the equation by 2:
$2X/2 = 1/2$
$X = 0.5$

3. $aX - bX = 1$
Factor X out of the term: $aX - bX$
$X (a - b) = 1$
Divide both sides of the equation by (a - b):
$X (a - b)/ (a - b) = 1/ (a - b)$
$X = 1/ (a - b)$

4. $aX - bX - cX = 2a - 2b - 2c$
Factor X out of the term: $aX - bX - cX$
$X (a - b - c) = 2(a - b - c)$

The Easiest Way to Understand Algebra

Divide both sides of the equation by (a - b - c):
X (a - b - c)/ (a - b - c)=2(a - b - c)/(a - b - c)
X = 2

5. 3abX - 5a = 3acX + 13a
Add 5a to both sides of the equation:
3abX - 5a + 5a = 3acX + 13a + 5a
3abX = 3acX + 18a
Subtract 3acX from both sides of the equation:
3abX - 3acX = 3acX + 18a - 3acX
3abX - 3acX = 18a
Factor 3aX out of the term: 3abX - 3acX
3aX (b - c) = 18a
Divide both sides of the equation by (b - c):
3aX (b - c)/ (b - c) = 18a/ (b - c)
3aX = 18a/ (b - c)
Divide both sides of the equation by 3a:
3aX/3a = [18a/(b - c)]/ 3a
X = 6/ (b - c)

6. aX - bX = ac - bc
Factor X out of the term: aX - bX
X (a - b) = c (a - b)
Divide both sides of the equation by (a - b):
X (a - b)/ (a - b) = c(a - b)/(a - b)
X = c

7. 9a - 4X = 5a - 2X
Add 4X to both sides of the equation:
9a - 4X + 4X = 5a - 2X + 4X
9a = 5a - 2X + 4X
9a = 5a + 2X
Subtract 5a from both sides of the equation:
9a - 5a = 5a + 2X - 5a
4a = 2X
2X = 4a
Divide both sides of the equation by 2:
2X/2 = 4a/2
X = 2a

8. X - aX = 2 - 2a
Factor X out of the term: X – aX
X (1 - a) = 2(1 - a)

39

Divide both sides of the equation by (1 - a):
X (1 - a)/ (1 - a) = 2(1 - a)/ (1 - a)
X = 2(1 - a)/ (1 - a)
X = 2

9. aX - bX = b - a
Factor X out of the term: aX - bX
X (a - b) = b - a
Factor out (-1) from the right side of the equation.
X (a - b) = (-b + a) (-1)
X (a - b) = (a - b) (-1)
Divide both sides of the equation by (a - b):
X (a - b)/ (a - b) = (-1) (a - b)/ (a - b)
X = -1(a - b)/ (a - b)
X = -1

Practice 5. Solutions.

1. 5aX - 5bX = 10b - 10a
Factor 5X out of the term: 5aX - 5bX
5X (a - b) = 10b - 10a
Factor -10 out of the term: 10b - 10a
5X (a - b) = -10(-b + a)
Divide both sides of the equation by (a - b):
5X (a - b)/ (a - b) = -10(a - b)/ (a - b)
5X = -10;
Divide both sides of the equation by 5.
5X/5 = -10/5
X = - 2

2. aX - bX - cX = c + b - a
Factor X out of the term: aX - bX - cX
X (a - b - c) = c + b – a
Factor out (-1) from the right side of the equation.
X (a - b - c) = (- c - b + a) (-1)
X (a - b - c) = (a - b - c) (-1)
Divide both sides of the equation by (a - b - c):
X (a - b - c)/ (a - b - c) = (a - b - c) (-1)/ (a - b - c)
X = -1

3. 2X - 3aX = 6a - 4
Factor X out of the term: 2X - 3aX
X (2 - 3a) = 2(3a - 2)

The Easiest Way to Understand Algebra

Factor out (-1) from the right side of the equation.
X (2 - 3a) = 2(-1)(- 3a + 2)
X (2 - 3a) = -2 (2 - 3a)
Divide both sides of the equation by (2 - 3a):
X (2 - 3a)/ (2 - 3a) = -2 (2 - 3a)/ (2 - 3a)
X = - 2

4. 3aX - 9bX = 27b - 9a
Factor X out of the term: 3aX - 9bX
3X (a - 3b) = 9(3b - a)
Factor out (-1) from the right side of the equation.
3X (a - 3b) = 9(-1)(a - 3b)
Divide both sides of the equation by (a - 3b):
3X (a - 3b)/ (a - 3b) = -9(a - 3b)/ (a - 3b)
3X = -9
Divide both sides of the equation by 3:
3X/3 = -9/3
X = - 3

5. 4bX - cX = 8c - 32b
Factor X out of the term: 4bX – cX
Factor 8 out of the term: 8c - 32b
X (4b - c) = 8(c - 4b)
Factor out (-1) from the right side of the equation.
X (4b - c) = 8(-1)(-c + 4b)
Divide both sides of the equation by (4b - c):
X (4b - c)/ (4b - c) = -8(-c + 4b)/ (4b - c)
X = - 8

6. abX - acX = ac - ab
Factor X out of the term: abX - acX
Factor a out of the term: ac - ab
aX (b - c) = a (c - b)
Factor out (-1) from the right side of the equation.
aX (b - c) = a(-1)(-c + b)
aX (b - c) = -a (b - c)
Divide both sides of the equation by (b - c):
aX (b - c)/ (b - c) = -a (b - c)/ (b - c)
Where (b-c) does not equal 0!
aX = -a
Divide both sides of the equation by a:
aX/a = -a/a
Where a is not equal to 0!

41

X = - 1

7. X/2 - aX = 1 - 2a
Multiply both sides of the equation by 2:
2(X/2 - aX) = 2 (1 - 2a)
X - 2aX = 2 (1 - 2a)
Factor X out of the term: X - 2aX
X (1 - 2a) = 2(1 - 2a)
Divide both sides of the equation by (1 - 2a):
X (1 - 2a)/ (1 -2a) = 2(1 - 2a)/ (1 -2a)
X = 2

8. aX/5 + 2a = 5a - 4aX
Subtract 2a from both sides of the equation:
aX/5 + 2a - 2a = 5a - 4aX - 2a
aX/5 = 5a - 4aX - 2a
Add 4aX to both sides of the equation:
aX/5 + 4aX = 5a - 4aX - 2a + 4aX
aX/5 + 4aX = 5a - 2a
aX/5 + 4aX = 3a
Multiply both sides of the equation by 5:
5(aX/5 + 4aX) =5 * 3a
aX + 20aX = 15a
21aX = 15a
Divide both sides of the equation by a:
21aX/a = 15a/a
Where a is not equal to 0!
21X = 15
Divide both sides of the equation by 21:
21X/21 = 15/21
X = 15/21
X = 5/7

Practice 6. Solutions.

1. X + Y = 1
   X - Y = - 5

We have two equations. Write one equation on top of the other and find their sum.

$$\begin{array}{r} X + Y = 1 \\ \underline{X - Y = -5} \\ 2X \quad\quad = -4 \end{array}$$

if $2X = -4$ then $2X / 2 = -4 / 2$ and
$X = -2$
To find Y, substitute X with (-2) in one of equations. $X + Y = 1$
$(-2) + Y = 1$
Then $(-2) + 2 + Y = 1 + 2$
$Y = 3$

2. $X + Y = 1$
   $2X - 2Y = 6$

In the first equation, we have positive Y, and in the second equation, we have negative 2Y.
To get rid of Y in an equation, we can multiply the first equation by 2.
We will get $2X + 2Y = 2$.
Now we can add the equations

$$+\begin{array}{r} 2X + 2Y = 2 \\ 2X - 2Y = 6 \\ \hline 4X + 0 = 8 \end{array}$$

$4X = 8$ then $X = 2$
Substitute X with 2 in the first equation:
$2+Y=1$ then $Y= 1-2$ then $Y = -1$.

3. $X + 2Y = 1$
   $2X + Y = -4$

In the first equation, we have X and in the second 2X.
To get rid of X, we can multiply the first equation by 2
$2(X + 2Y) = 2 *1$ then we get $2X + 4Y = 2$.
Now we subtract the second equation from the first equations.

$$\begin{array}{r} 2X + 4Y = 2 \\ 2X + Y = -4 \\ \hline 0 + 3Y = 6 \end{array}$$

We get 6 because $2 - (-4) = 6$
Then $3Y = 6$ and $Y=2$. Substitute Y with 2 in the first equation.
$X + 4=1$
Subtract 4 from both sides of the equation
$X + 4 - 4 = 1 - 4$
then $X = -3$
Substitute X and Y with their values in the first equation:
$(-3) + 2 * 2 = 1$ then $-3 + 4 = 1$ and $1=1$

4. $2X + Y = 5$
   $X + 3Y = 0$
Multiply the second equation by 2.
$2(X + 3Y) = 2 * 0$ then $2X + 6Y = 0$
Subtract the new second equation from the first one.

$$\begin{array}{r} 2X + Y = 5 \\ -\underline{2X + 6Y = 0} \\ 0 - 5Y = 5 \end{array}$$

Then $-5Y/-5 = 5/-5$ and $Y = -1$
Substitute Y in the first equation with -1
Then $2X - 1 = 5$
Add 1 to both sides of the equation.
$2X - 1 + 1 = 5 + 1$
$2X = 6$ and $X = 3$.
Substitute X and Y with their values in the first equation:
$2 * 3 - 1 = 5$
$5 = 5$

5. $3X - Y = 5$
   $4X + 2Y = 10$
Multiply the first equation by 2 then we will get
$6X - 2Y = 10$.
Add the second equation to the first one

$$\begin{array}{r} 6X - 2Y = 10 \\ \underline{4X + 2Y = 10} \\ 10X + 0 = 20 \end{array}$$

Then $X = 20/10 = 2$
Substitute X in the first equation:
$3*2 - Y = 5$
$6 - Y = 5$. Subtract 6 from both sides of the equation and you get:
$-Y = -1$
Multiply both sides of the equation with -1 and you get:
$Y = 1$
Substitute X and Y in the second equation:
$4*2 + 2*1 = 10$ and $8 + 2 = 10$ $10 = 10$
Substitute X and Y in the first equation:
$3*2 - 1 = 5$
$5 = 5$

6. $4X + 2Y = 10$
   $4X - 2Y = 6$

The Easiest Way to Understand Algebra

Now add the equations:
$$\begin{array}{r}4X + 2Y = 10\\ +\ 4X - 2Y = 6\\ \hline 8X + 0 = 16\end{array}$$

$8X/8 = 16/8$ and $X = 2$
Substitute X in the first equation.
$4 * 2 + 2Y = 10$ You get:
$8 + 2Y = 10$
Subtract 8 from both sides of the equation:
$8 - 8 + 2Y = 10 - 8$
$2Y = 2$
$Y = 1$
Substitute X and Y in the second equation:
$4 * 2 - 2 * 1 = 6$
$8 - 2 = 6$
$6 = 6$
Substitute X and Y in the first equation:
$4 * 2 + 2 * 1 = 10$
$8 + 2 = 10$
$10 = 10$

7. $X - Y = 1.5$
   $7X + 2Y = 6$
Multiply the first equation by 2 and get:
$2X - 2Y = 3$
Sum the equations:

$$\begin{array}{r}2X - 2Y = 1.5\\ +\ 7X + 2Y = 6\\ \hline 9X + 0 = 9\end{array}$$

$9X = 9$ then $X = 1$
Substitute X in the first equation:
$1 - Y = 1.5$
$1 - 1 - Y = 1.5 - 1$
Then $-Y = 0.5$ and $Y = -0.5$
Substitute X and Y in the first equation:
$1 - (-0.5) = 1.5$ and $1 + 0.5 = 1.5$
$1.5 = 1.5$
Substitute X and Y in the second equation:
$7*1 + 2(-0.5) = 7 - 1 = 6$ and $6 = 6$

8. $3X + Y = 9$

X - 3Y = 3
Multiply the first equation by 3 and get:
9X + 3Y = 27
Sum the equations:

$$+\begin{array}{r} 9X + 3Y = 27 \\ X - 3Y = 3 \\ \hline 10X + 0 = 30 \end{array}$$

X = 30/10 = 3
Substitute X in the first equation and find Y:
3*3 + Y = 9
Y = 9-9 = 0
Substitute X and Y in the first equation:
3*3 + 0 = 9 and 9=9
Substitute X and Y in the second equation:
3 - 3*0 = 3 and 3 = 3

9. 5X - 2Y = - 7
   X + 3Y = 2
Multiply each part of the second equation by -5 and get:
-5X - 15Y = -10
Sum the equations:

$$+\begin{array}{r} 5X - 2Y = -7 \\ -5X - 15Y = -10 \\ \hline 0 - 17Y = -17 \end{array}$$

-17Y = -17 then Y = 1
Substitute Y in the second equation and find X:
X + 3*1 = 2
then X - 3 + 3 = 2 - 3
X = - 1
Substitute X and Y in the first equation:
5(-1) - 2*1 = -7 then
-5 - 2 = -7 and -7 = -7
Substitute X and Y in the second equation:
-1 + 3*1 = 2 then -1 + 3 = 2 and 2 = 2

10. 9X + 3Y = 12
    X - 2Y = - 1
Multiply the second equation by 9 and get
9X - 18Y = -9

Subtract the second equation from the first one:
$$\begin{array}{r} 9X + 3Y = 12 \\ -\underline{9X - 18Y = -9} \\ 0 + 21Y = 21 \end{array}$$

We get 21Y because 3Y - (-18Y) = 21Y   Y = 1
the same way we get 21 (12 -(-9) = 21
Substitute Y in the second equation and find X:
X - 2*1 = -1 then
X = -1 + 2 = 1
Substitute X and Y in the first equation:
9*1 + 3*1 = 12 12 = 12
Substitute X and Y in the second equation:
1- 2*1 = -1 then 1 - 2 = -1 and -1 = -1

Practice 6. Graph solutions.

To solve simultaneous equations graphically, we have to plot a graph for each equation. The intersection point of these two lines gives us the solution.
1. X + Y = 1
   X - Y = - 5

Find 2 points to draw the first line: X + Y = 1
Modify the equation to a general form Y = aX + b
Subtract X from both sides of the equation.
X - X + Y = 1 − X
Y = 1 - X
X = 0 Y (0) = 1 - 0 = 1
X = 5 Y (5) = 1 - 5 = - 4

Find 2 points to draw the second line: X - Y = - 5
Modify the equation to a general form Y = aX + b
Subtract X from both sides of the equation.
X - Y - X = - 5 - X
- Y = - 5 - X
Multiply both sides by -1 and you get:
Y = 5 + X
X = 0 Y (0) = 5 + 0 = 5
X = -5 Y (-5) = 5 - 5 = 0

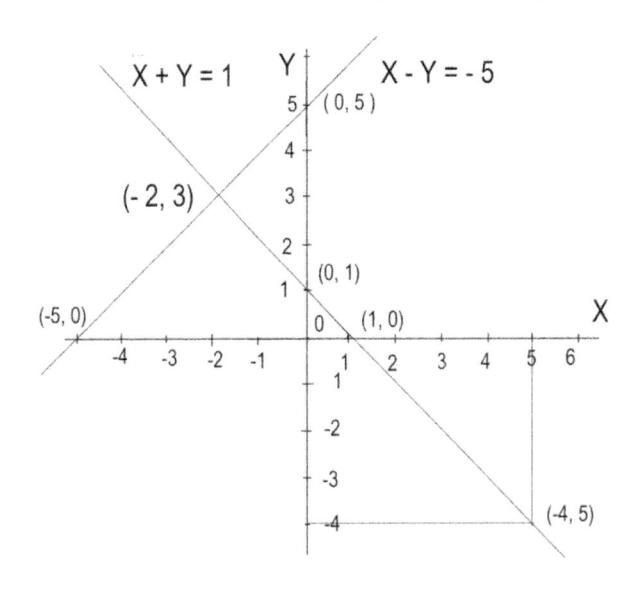

Graph 1. The point of the intersection: X=-2, Y= 3.

2. X + Y = 1
   2X - 2Y = 6

Y = 1 - X
The first line points:
X = 0; Y (0) = 1
X = 5; Y (5) = 1 - 5 = - 4

The second line points:
-2Y = -2X + 6
Divide both sides by -2
Y = X - 3
X = 0; Y (0) = -3
X = 3; Y (3) = 3 - 3 =0

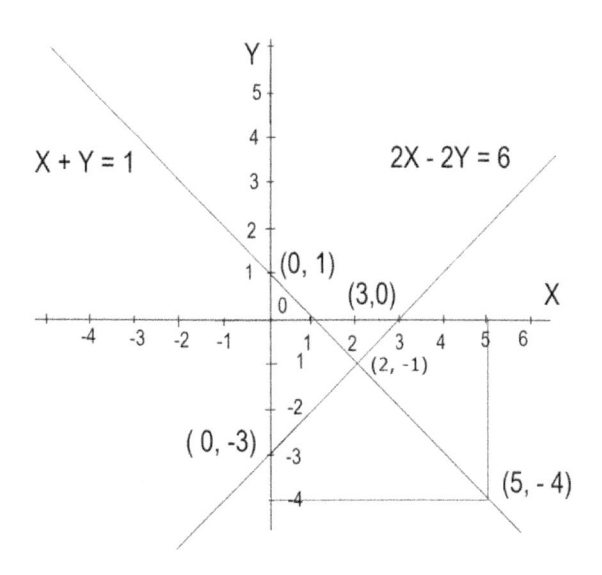

Graph 2. The point of the intersection: X= 2, Y=-1.

3. X + 2Y = 1
   2X + Y = - 4

The first line points: X + 2Y = 1

2Y = 1 - X
Y = (1 - X)/2
X = 5; Y (5) = (1 - 5)/2 = - 2
X = -5; Y (-5) = (1 -(- 5)/2 = 3

The second line points: 2X + Y = - 4

Y = - 4 - 2X
X = 0; Y (0) = - 4
X = -4; Y (-4) = - 4 - 2 (- 4) = - 4 +8 = 4

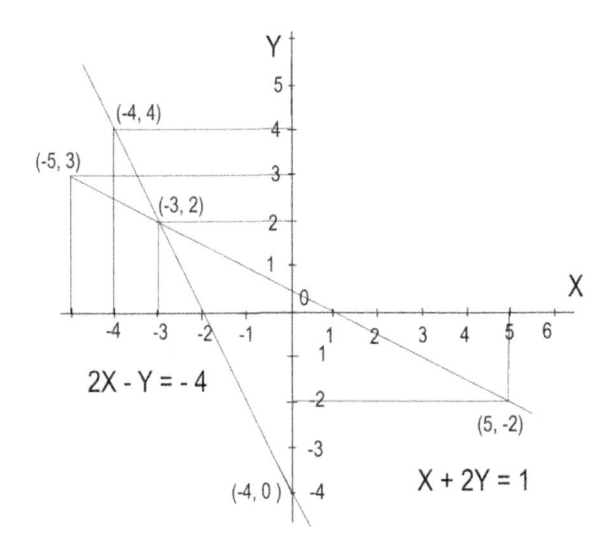

Graph 3. The point of the intersection: X=-3, Y=2.

4. 2X + Y = 5
   X + 3Y = 0

The first line points: 2X + Y = 5

2X - 2X + Y = 5 - 2X
Y = 5 - 2X
X = 0; Y (0) = 5
X = 3; Y (3) = 5 - 2*3 = -1

The second line points: X + 3Y = 0

X - X + 3Y = - X
3Y = - X
Y = - X / 3
X = 6; Y (6) = -2
X = -3; Y (-3) = 1

The Easiest Way to Understand Algebra

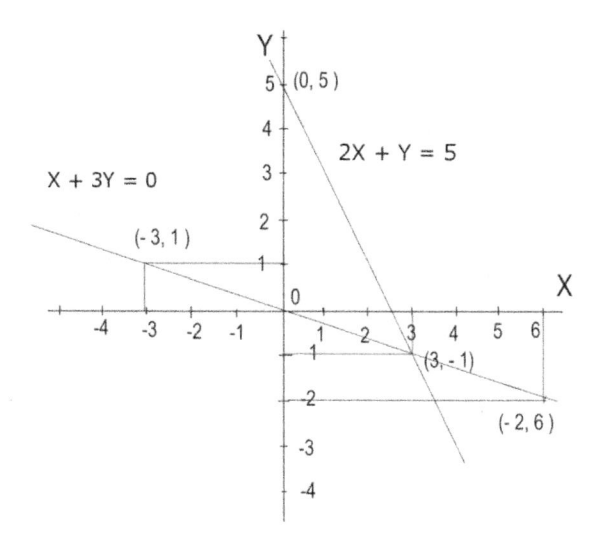

```
              Y
            5  (0, 5)
            4
                        2X + Y = 5
X + 3Y = 0  3
            2
   (-3, 1)
            1
                                        X
            0
-4  -3  -2  -1   1   2   3   4   5   6
                                (3, -1)
           -2
                                  (-2, 6)
           -3
           -4
```

Graph 4. The point of the intersection: X=3, Y=-1.

5. 3X - Y = 5
4X + 2Y = 10

The first line points: 3X - Y = 5

3X - 3X - Y = 5 - 3X
-Y = 5 - 3X
Y = 3X - 5
X = 0; Y (0) = -5
X = 2; Y (2) = 3*2 -5 =1

The second line points: 4X + 2Y = 10

Subtract -4X from both sides of the equation.
4X - 4X + 2Y= 10 - 4X
2Y = 10 - 4X
Divide both sides of the equation by 2.
2Y/2 = 10/2 - 4X/2
Y = 5 - 2X
X = 0; Y = 5
X = 4; Y = 5 -2*4= -3

$4X + 2Y = 10$ ⋯ $3X - Y = 5$

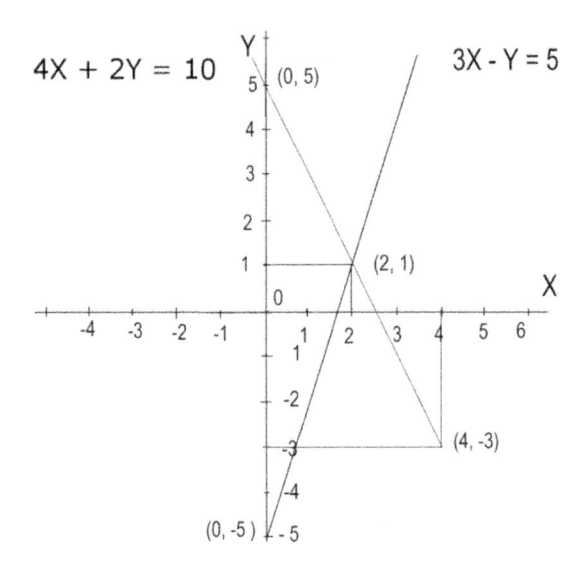

Graph 5. The point of the intersection: X=2, Y=1.

6. $4X + 2Y = 10$
   $4X - 2Y = 6$

The first line points: $4X + 2Y = 10$

Subtract 4X from both sides of the equation.
$4X - 4X + 2Y = 10 - 4X$
$2Y = 10 - 4X$
Divide both sides by 2.
$2Y/2 = 10/2 - 4X/2$
$Y = 5 - 2X$
$X = 0; Y (0) = 5 - 2*0 = 5$
$X = 5; Y (5) = 5 - 2 * 5 = 5 – 10 = -5$

The second line points: $4X - 2Y = 6$

$4X - 4X - 2Y = 6 - 4X$
$-2Y = 6 – 4X$
$-2Y/2 = 6/2 - 4X/2$
$- Y = 3 – 2X$
$(-1)(-Y) = (-1)(3 – 2X)$
$Y = 2X - 3$
$X = 0; Y (0) = 2*0 - 3 = -3$
$X = 4; Y (4) = 2*4 - 3 = 5$

$4X + 2Y = 10$

$4X - 2Y = 6$

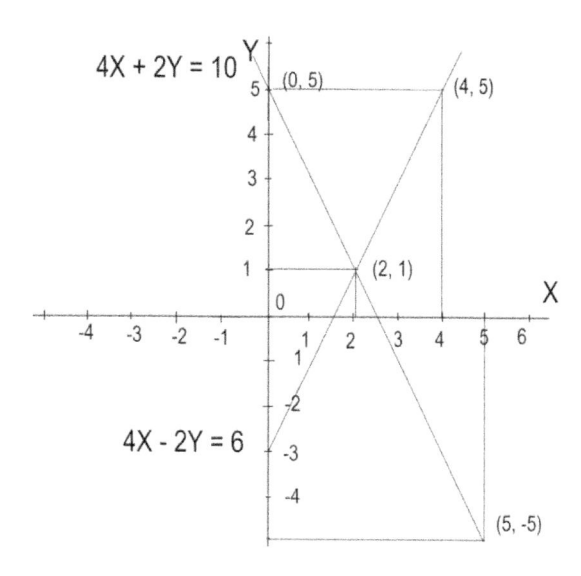

Points shown on graph: (0, 5), (4, 5), (2, 1), (5, -5)

Graph 6. The point of the intersection: X= 2, Y=1.

7. $X - Y = 1.5$
   $7X + 2Y = 6$

The first line points: $X - Y = 1.5$
Subtract X from both sides of the equation.
$X - X - Y = 1.5 - X$
$- Y = 1.5 - X$
Multiply both sides by -1
$(-1)(-Y) = (-1)(1.5 - X)$
$Y = (-1.5 + X)$
$Y = X - 1.5$
$X = -2$; $Y(-2) = -2 - 1.5 = -3.5$
$X = 5$; $Y(5) = 5 - 1.5 = 3.5$
The second line points: $7X + 2Y = 6$
Subtract 7X from both sides of equation.
$7X - 7X + 2Y = 6 - 7X$
$2Y = 6 - 7X$
Divide both sides of the equation by 2
$2Y/2 = 6/2 - 7X/2$
$Y = 3 - 7X/2$
$X = 0$; $Y(0) = 3$
$X = 2$; $Y(2) = 3 - 7*2/2 = 3 - 7 = -4$

7X + 2Y = 6   X - Y = 1.5

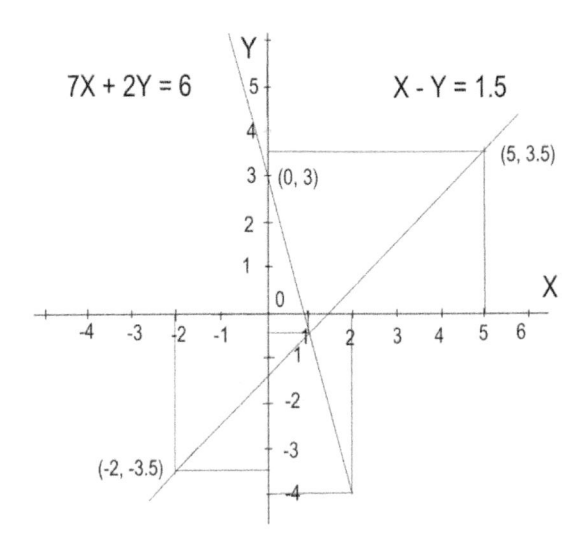

(0, 3)
(5, 3.5)
(-2, -3.5)

Graph 7. The point of the intersection: X=1, Y=-0.5.

8. 3X + Y = 9
   X - 3Y = 3

The first line points: 3X + Y = 9

Subtract 3X from both sides of the equation.
3X - 3X + Y = 9 - 3X
Y = 9 - 3X
X = 4; Y (4) =9 -3*4=9 – 12 =- 3
X= 2; Y (2) =9 - 3*2= 3

The second line points: X - 3Y = 3

Subtract X from both sides of the equation.
X - X - 3Y = 3 - X
- 3Y = 3 - X multiplied by -1
3Y = X - 3
Y = (X -3)/3
X = 0; Y (0) = - 1
X = - 3; Y (3) = (- 3 – 3)/3 = -2

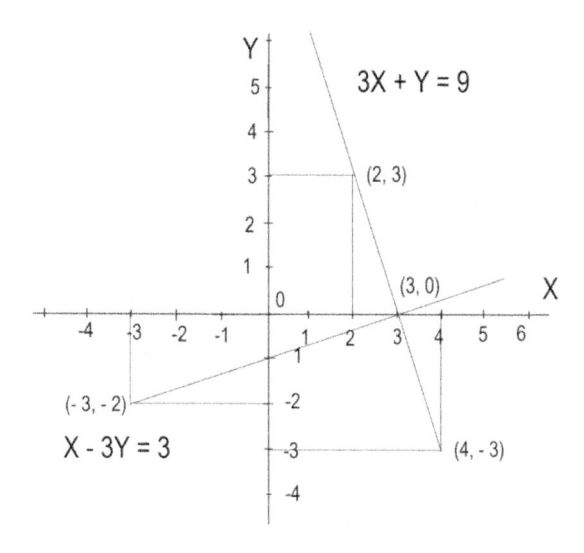

Graph 8. The point of the intersection: X=3, Y=0.

9. $5X - 2Y = -7$
$X + 3Y = 2$

The first line points: $5X - 2Y = -7$

Subtract 5X from both sides of the equation.
$5X - 5X - 2Y = -7 - 5X$
$-2Y = -7 - 5X$
Multiply both sides of the equation by -1 and you get:
$2Y = 7 + 5X$
Divide both sides of the equation by 2 and you get:
$Y = (7 + 5X)/2$
$X = 0; Y(0) = 7/2 = 3.5$
$X = 1; Y(1) = (7 + 5)/2 = 6$

The second line points: $X + 3Y = 2$

$X - X + 3Y = 2 - X$
$3Y = 2 - X$
$Y = (2 - X)/3$
$X = -4; Y(-4) = (2 - (-4))/3 = 2$
$X = 5; Y(5) = (2 - 5)/3 = -1$

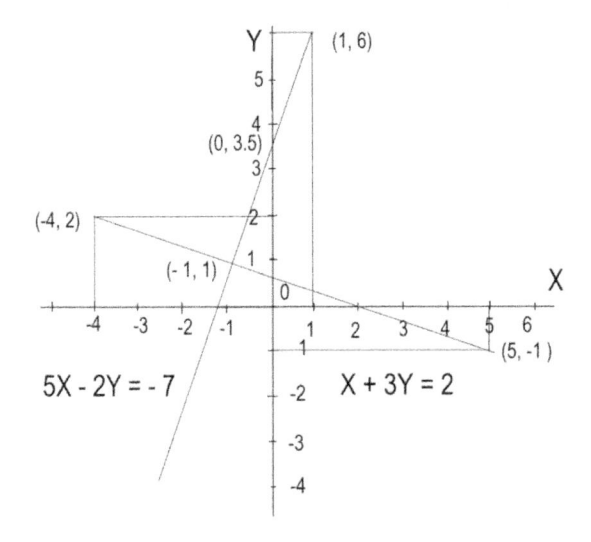

[Graph showing two lines intersecting, with labeled points (1, 6), (0, 3.5), (-4, 2), (-1, 1), (5, -1), and equations 5X - 2Y = -7 and X + 3Y = 2]

Graph 9. The point of the intersection: X= -1, Y=1.

10. 9X + 3Y = 12
X - 2Y = - 1

The first line points: 9X + 3Y = 12

Subtract 9X from both sides of the equation:
9X - 9X + 3Y = 12 – 9X
3Y = 12 - 9X
Divide both sides of the equation by 3 and you get:
Y = 4 - 3X
X = 0; Y (0) = 4
X = 3; Y (3) = 4 - 3(3) = 4 - 9 = -5

The second line points: X - 2Y = - 1

Subtract X from both sides of the equation.
X - X - 2Y = -1 - X
-2Y = - 1 - X
Divide both sides of the equation by - 2.
-2Y/-2 = (-1 - X)/-2
Y = (1 + X)/2
X = -5; Y (-5) = (1 - 5)/2= - 4/2 = - 2
X = 5; Y (3) = (1 + 5)/2=3

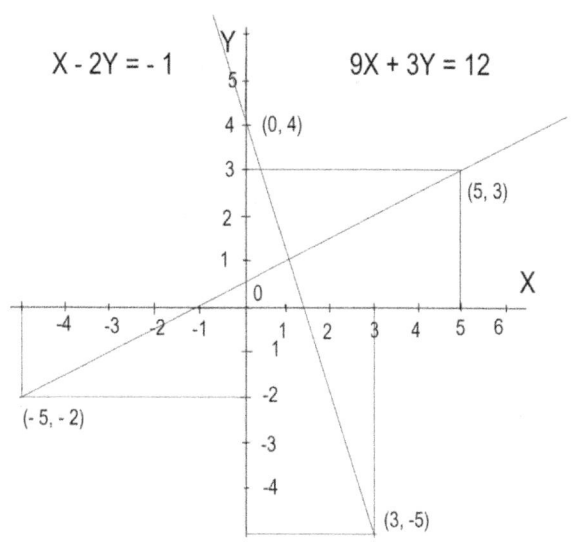

```
         Y
X - 2Y = -1  |        9X + 3Y = 12
         5
         4  (0, 4)
         3
         2            (5, 3)
         1
         0                        X
-4 -3 -2 -1   1  2  3  4  5  6
         -1
         -2
(-5, -2)
         -3
         -4
              (3, -5)
```

Graph 10. The point of the intersection: X=1, Y=1.

Practice 7. Solutions.

1. $3X^2 - 75 = 0$
Divide each part of equation by 3 and get:
$X^2 - 25 = 0$
Factor the resulting equation:
$(X - 5)(X + 5) = 0$
then $X - 5 = 0$ or $X = 5$
$X + 5 = 0$ and $X = -5$.
The answer is X may be 5 or -5.
Let us check if $X^2 - 25 = 0$
$(-5)(-5) = 25$ and $5 * 5 = 25$ We are correct.

2. $2X^2 - 9X + 4 = 0$
Factor the equation and get:
$(2X - 1)(X - 4) = 0$ To check
multiply $(2X - 1)(X - 4)$
and you will get the same equation:
$2X^2 - 8X - 1X + 4 =$
$= 2X^2 - 9X + 4$
Then $2X - 1 = 0$ and $X = 0.5$
Find the second X value:
$X - 4 = 0$
$X = 4$

Check X=0.5:
2 * (0.5) (0.5) - 9 (0.5) + 4 =0
2* 0.25 - 4.5 + 4 =0
and 0.5 - 4.5 + 4 =0
4.5 - 4.5 =0.
Check X = 4
2 * 4 * 4 - 9 * 4 + 4 = 0
then 32 - 36 + 4 = 0 and - 4 + 4 =0

3. $3X^2 -5X -2=0$
Factor the equation:
(3X + 1) (X -2) =0
Check 3X * X - 6X + X - 2 =
=$3X^2$ -5X -2=0 We are correct.
3X + 1 = 0 then 3X = -1 and X = -1/3
X - 2 = 0 then X = 2
Let us substitute the first X in the equation:
3 * (-1/3) (-1/3) - 5 (-1/3) -2 =
=3/9 + 5/3 -2 =
=3/9 + 15/9 -18/9 = 18/9 - 18/9 =0
Let us substitute the second X in the equation:
3*2*2 - 5*2 - 2 =0 then 12 -10 -2=0

4. $4X^2 -13X +3=0$ Factor the equation
(4X - 1) (X - 3) = 0
To check multiply: (4X - 1) (X - 3)
and you will get the same equation:
4X*X - 12X - X + 3 =
=$4X^2$ -13X +3
4X -1 = 0 then X = 0.25
X - 3 = 0 then X = 3
Let us substitute the first X in the equation:
4*0.25*0.25 - 13*0.25 +3 =
=0.25 - 3.25 + 3= 0
We are correct.
Let us substitute the second X in the equation:
4*3*3 -13*3 +3 =
=36 - 39 +3 =0
We are correct.

5. $7X^2 - 29X +4=0$ Factor the equation:
(7X - 1) (X - 4) =0
Check: (7X - 1) (X - 4) =

The Easiest Way to Understand Algebra

=7X*X - 28X - X + 4 =
7X^2 - 29X + 4
7X -1 = 0 then
X = 1/7; X-4 = 0
then X = 4
Let us substitute the first X in the equation:
7 (1/7) (1/7) - 29(1/7) + 4 =0
7/49 - 29/7 + 4 =
=7/49 - 29*7/49 + 4*49/49=
7/49 - 203/49 + 196/49=
203/49 - 203/49 =0
Let us substitute the second X in the equation
7*4*4 - 29*4 + 4 =
=112 - 116 + 4=0
We are correct.

6. 5X^2 -28X +15=0 Factor the equation:
(5X - 3) (X - 5) =0
Check: (5X - 3) (X - 5) =
=5X*X - 25X - 3X + 5 =
5X^2 -28X +15
5X -3 =0 then X =3/5
X - 5 = 0 then
X = 5
Let us substitute the first X in the equation:
5*(3/5) (3/5) - 28(3/5) + 15 =
=5 (9/25) - 28 (3/5) + 15 =
=9/5 - 84/5 + 75/5 =
-75/5 + 75/5 =0
We are correct.
Let us substitute the second X in the equation:
5* 5*5 -28*5 +15 =
=125 - 140 + 15=0
We are correct.

7. 18X^2 +12X -6=0 Factor the equation
(3X + 3) (6X - 2)
Check: (3X + 3) (6X - 2) =
18X*X - 6X + 18X - 6 =
18X*X + 12X - 6 =
18X^2 +12X -6
3X + 3 =0 then
X = -1

59

6X -2 = 0
then X = 1/3
Let us substitute the first X in the equation:
18 * (-1) (-1) + 12(-1) - 6 =
=18 - 12 - 6 =0
Let us substitute the second X in the equation:
18 (1/3) (1/3) + 12(1/3) - 6 =
=18/9 + 12/3 - 6 =
6/3 + 12/3 - 18/3 =
18/3 - 18/3 =0
We are correct.

8. $3X^2$ -12X + 9=0 Factor the equation
(3X - 3) (X -3) = 0
Check: (3X - 3) (X -3) =
=3X * X - 9X - 3X + 9 =
3X * X - 12X + 9 =
3X-3 = 0 then X =1
X - 3 = 0 then X=3
Let us substitute the first X in the equation:
3* 1*1 -12*1 +9 =
=3 - 12 + 9 =0
We are correct.
Let us substitute the second X in the equation:
3*3*3 -12*3 + 9 =
=27 - 36 +9 = 0
We are correct.

9. $24X^2$ +55X -24=0 Factor the equation:
(8X - 3) (3X + 8)
Check: (8X - 3) (3X + 8) =
=24X * X + 64X - 9X - 24 =
=24X * X + 55X - 24 =
=$24X^2$ + 55X -24
8X -3 = 0 then
X = 3/8
3X + 8 = 0
then X = -8/3
Let us substitute the first X in the equation:
24* 3/8*3/8 + 55*3/8 -24=
= 24* 9/64 + 165/8 -24 =
3*9/8 + 165/8 - 24*8/8=
27/8 + 165/8 -192/8 =

192/8 - 192/8 = 0 We are correct.
Let us substitute the second X in the equation:
24 * (-8/3) (-8/3) + 55(-8/3) - 24=
=24 (64/9) - 440/3 - 24 =
512/3 - 440/3 - 24* 3/3 =
=512/3 - 440/3 - 72/3 = 0
We are correct.

10. $12X^2$ -45X -12=0 Factor the equation:
(4X + 1) (3X - 12) =0
Why did I not choose
(4X + 3) (3X - 4)?
4X * 3X =$12X^2$ and 3 * (-4) = -12.
I did not choose (4X + 3) (3X - 4) and have chosen
(4X + 1) (3X - 12) instead because in the equation we have
-45X as a second member.
(4X + 3) (3X - 4) can give us -16X (4X * -4) and 9X (3*3X)
We cannot get -45X from (4X + 3) (3X - 4)
I have chosen (4X + 1) (3X - 12)
because 4X * 12 gives us 48X and it is close to 45X.
Let us check:
(4X + 1) (3X - 12) =
=12X*X - 48X + 3X - 12 =
12X*X - 45X -12 =
=$12X^2$ -45X -12
4X + 1 = 0
Then X = - 0.25
3X - 12 = 0
Then X = 12/3=4
Let us substitute the first X in the equation:
12*(-0.25) (-0.25) - 45(0.25) - 12 =
0.75 - 45*(-0.25) - 12 =
=0.75 + 11.25 - 12 =
=12 - 12 =0
We are correct.
Let us substitute the second X in the equation:
12* 4*4 - 45*4 -12 =
=192 - 180 - 12=0
We are correct.

11. $8X^2$ -28X +12=0 Factor the equation:
(4X - 2) (2X - 6)
Check: (4X - 2) (2X - 6) =

4X*2X - 24X - 4X + 12 =
=8X^2 -28X +12
4X - 2 = 0 then X = 0.5
2X-6 =0 then X =3
Let us substitute the first X in the equation:
8*(0.5) (0.5) - 28*(0.5) + 12=
=2 - 14 + 12 =0
We are correct
Let us substitute the second X in the equation:
8*3*3 -28*3 +12 =
=72 - 84 + 12=0
We are correct.

12. 12X^2 - 48=0
Divide the equation by 12 and get
X^2 - 4=0
Factor the equation: (X - 4) (X + 0) =0
Check: (X - 4) (X + 0) =
=X*X + 0 + 4X - 0) =
=X^2 - 4=0
X - 4 = 0 then X = 4
X + 0 = 0 then X =0
Let us substitute the first X in the equation:
12*4*4 - 48*4 =0 we are correct.
Let us substitute the second X in the equation:
12*0 - 48*0 = 0 we are correct.

13. 2X^2 - 11X + 9=0 Factor the equation:
(2X - 9) (X - 1)
Check: (2X - 9) (X - 1) =
=2X*X - 2X - 9X + 9 =
2X^2 - 11X + 9
(2X - 9) = 0 then X =4.5
X - 1 = 0 then X =1
Let us substitute the first X in the equation:
2* 4.5 * 4.5 - 11*4.5 +9 =
= 40.5 - 49.5 + 9=0
We are correct.
Let us substitute the second X in the equation:
2 * 1*1 - 11*1 +9 =
= 2 - 11 + 9 =0 We are correct.

14. 14X^2 - 27X + 9 = 0 Factor the equation:

The Easiest Way to Understand Algebra

(2X - 3) (7X - 3)
Check: (2X - 3) (7X - 3) =
2X*7X - 21X - 6X + 9 =
=14X*X - 27X +9 =
=14X^2 - 27X +9
2X - 3 = 0 then X = 1.5
7X -3 = 0 then X = 3/7
Let us substitute the first X in the equation:
14*1.5*1.5 - 27*1.5 + 9 =
31.5 - 40.5 + 9 = 0
We are correct
Let us substitute the second X in the equation:
14*3/7*3/7 - 27*3/7 +9=
=14*9/49 - 27*3/7 +9=
=2*9/7 - 27*3/7 +9=
18/7 - 81/7 + 63/7 =
81/7 - 81/7 =0
We are correct.

15.5X^2 - 37X + 14=0 Factor the equation:
(5X - 2) (X - 7)
Check: (5X - 2) (X - 7) =
=5*X*X - 35X - 2X + 14 =
=5*X*X - 37X + 14 =
=5X^2 - 37X +14
5X - 2 = 0 then X = 0.4
X - 7 = 0 then X =7
Let us substitute the first X in the equation:
5*0.4*0.4 - 37*0.4 + 14 =
=0.8 - 14.8 + 14=0
We are correct.
Let us substitute the second X in the equation:
5 *7*7 - 37*7 +14 =
=245 -259 +14 =0
We are correct.

Practice 8 solutions

1. $\log_{10} 3x = -1$

Represent -1 as the decimal logarithm of 10 raised by the power of -1.

$\log_{10} 3x = \log_{10} 10^{-1}$

Then $3x = 10^{-1}$

3x / 3 = (1 / 10) /3

X = 1/30

2. $\log_{10} 3x = \log_{10} 6x + 2$

Move the term with x to the left.

$\log_{10} 3x - \log_{10} 6x = 2$

Per the Quotient rule

$\log_{10} 3x - \log_{10} 6x = \log_{10} \frac{3x}{6x}$ then

$\log_{10} \frac{3x}{6x} = 2$

Represent 2 as the decimal logarithm of 100.

$\log_{10} \frac{3x}{6x} = \log_{10} 100$

Then 3x / 6x = 100

0.5x = 100, x = 200

3. $\log_{10} 10^{x+10} = \log_{10} 100$

After applying the power rule of logarithms, we get;

$(x + 10) \log_{10} 10 = \log_{10} 100$

Divide both sides by logarithm of 10.

$$\frac{(x + 10) \log_{10} 10}{\log_{10} 10} = \frac{\log_{10} 100}{\log_{10} 10}$$

$$x + 10 = \frac{\log_{10} 100}{\log_{10} 10}$$

X + 10 = 2 / 1 = 2

X = 2 − 10 = − 8

4. $\log_{10}(x^2 - 2x + 2) = 1$

You can substitute 1 with log10 because 10 raised by the power of 1 equals 10.

$\log_{10}(x^2 - 2x + 2) = \log_{10} 10$

Then

$x^2 - 2x + 2 = 10$

Move 1 to the right side of the equation.

$x^2 - 2x - 8 = 0$

To solve a quadratic equation, use the following formula.

$ax^2 + bx + c = 0$

$$X = \frac{-b \pm \sqrt{b^2 - 4ac}}{2a}$$

Substitute the a, b and c with the numbers from your equation.

$$X = \frac{-(-2) \pm \sqrt{2^2 - 4 \bullet (-8)}}{2 \bullet 1}$$

2 raised by power of 2 = 4.

Minus 4 multiplied by minus 8 equals plus 36.
 The square root of 36 equals 6.

X = (2 + 6)/2 = 4   or X = (2 – 6)/2= - 2

Substitute it to the initial equation to check that the answers are correct.

$$\log_{10}(x^2 - 2x + 2) = 1$$

$$\log_{10}(4^2 - 2 \cdot 4 + 2) = 1$$

$$\log_{10} 10 = 1$$

The answer X = 4 is correct.

Let's check the second answer: x= - 2.

$$\log_{10}((-2)^2 - 2 \cdot (-2) + 2) = 1$$

$$\log_{10} 10 = 1$$

The answer X = - 2 is correct.

5. $\log_{10} 1/3x = 2$

Remember the rule : $\log 1/N = -\log N$ then

$-\log_{10} 3x = 2$

10 raised by the power of 2 equals 100 then

$-\log_{10} 3x = \log_{10} 100$

Multiply each side of the equation by – 1.

$\log_{10} 3x = -\log_{10} 100$

$\log_{10} 3x = \log_{10} 1/100$

3x = 1/100 and x = 1/300

There is a different way of solving this equation.

$\log_{10} 1/3x = \log_{10} 100$

Then 1/3x = 100

1/3x * 3x = 100 * 3x

1 = 300x

X = 1/300

6. $\log_{10} 40x + 1 = \log_{10} 4x + 3x$

Move the term with x to the left and numbers without x to the right.

$\log_{10} 40x - \log_{10} 4x - 3x = -1$

Per the Quotient rule

$$\log a/b = \log a - \log b$$

then

$$\log_{10} \frac{40x}{4x} - 3x = -1$$

40x/4x = 10 then

$$\log_{10} 10 - 3x = -1$$

Log 10 = 1 then

$$1 \qquad\qquad -3x = -1$$

-3x = - 2

X = 2/3

7. $\log_{10} 4x = 2 - \log_{10} x$

Move the term with x to the left and numbers without x to the right.

$$\log_{10} 4x + \log_{10} x = 2$$

Per the Product rule

$$\log ab = \log a + \log b$$

Then

$$\log_{10} 4x \cdot x = 2$$

Since 10 raised by the power of 2 equals 100 then

$$\log_{10} 4x^2 = \log_{10} 100$$

$$4x^2 = 100$$

Divide both sides by 4

$x^2 = 25$

X = 5 or X = -5

8. $\log_{10} x^{(x+1)} = x + 1$

Per the power rule

$\log N^m = m \log N$ then

$(x+1)\log_{10} x = x + 1$

Divide both sides by (x + 1)

$$\frac{(x+1)\log_{10} x}{(x+1)} = \frac{x+1}{(x+1)}$$

$\log_{10} x = 1$

$\log_{10} x = \log_{10} 10$

X = 10

9. $\log_{10} 2x^2 = 2 + \log_{10} x$

Move the term with x to the left

$\log_{10} 2x^2 - \log_{10} x = 2$

Per the Quotient rule:

$\log_{10} \frac{2x^2}{x} = 2$

10 raised by the power of 2 equals 100.

$$\log_{10}\frac{2x^2}{x} = \log_{10}100$$

Divide 2X^2 by X.

$$\log_{10}2x = \log_{10}100$$

Then 2x = 100 and x = 50.

10. $\log_{10}x = 3 - \log_{10}x^2$

Move the term with X to the left

$$\log_{10}x + \log_{10}x^2 = 3$$

Per the Product rule

$$\log_{10}x \cdot x^2 = 3$$

10 raised by the power of 3 equals 1000 then

$$\log_{10}x \cdot x^2 = \log_{10}1000$$

$x \cdot x^2 = x^3$ then

$X^3 = 1000$

X = 10

Thanks for reading! If you found this book useful, I'd be very grateful if you'd post a short review on Amazon. Your support does make a difference. Your feedback helps me to make this book even better.

Would you be open to sharing how this book helped you? Your words will serve other readers to benefit from this work. If you'd prefer not to, that's all good as well. Thanks again for your support!

My eBooks on Amazon.com

Roy Sawyer

Chemistry for Students and Parents
The Easiest Way To Understand Algebra
Geometry For Students and Parent
How to Create and Store Your Passwords

Sergey Skudaev

C++ Programming By Examples
PHP Programming for Beginners
Learn SQL by Examples
The Ultimate eBook Creator
Critical Computer programming Concept

ABOUT THE AUTHOR

Roy Sawyer currently living in Florida.

He obtained a master's degree in biology from a foreign University, where he specialized in neuropsychology. He also has a degree in Computer Science from Borough of Manhattan Community College, which he attained after moving to the US.

Since then, Roy has been working as a software quality engineer and web developer for a computer company in Florida.

He has more than ten years of teaching experience and a long-standing interest in new computer technologies, psychology, and brain physiology.

When he has some time to relax, Roy enjoys swimming in the ocean off the Florida coast or going for walks with his dog. He also enjoys traveling, particularly in the USA.

You can contact Roy at support@learn-coding.today

Printed in Great Britain
by Amazon